How does order emerge in social systems?

Applied Chaos Theory and Complex System Theory to Crime Prevention

Santiago Roel R

June 2010

Santiago Roel R

ISBN-13: 978-1511818605

ISBN-10: 1511818603

The Story

Sonora, Mexico. August 2008

Sonora is Mexico's second largest state, characterized by a large desert and a dispersed population. It borders with Arizona to the North, the Sea of Cortes to the West and three Mexican states that are home to different drug cartels: Chihuahua, Sinaloa and Baja California Norte. And yet, Sonora has radically reduced crime in the past year: Rape has decreased by 26%, Family Violence 34%, Assault 17%, House burglary 22%, and Robbery has reduced 30%.

These decreases are impressive if we compare Sonora to the rest of Mexico where violent crime is on the rise and has forced the President to hold an extraordinary meeting with his Security Council and all of Mexico's state governors. Yet, Sonora is not complacent with its achievements, it is way ahead doing something radically different.

Two years ago, Governor Eduardo Bours named Francisco Figueroa as Secretary of Security - one of his closest advisors. Figueroa sketched out a plan and invited me over to take a look in November 2006.

I didn't like his plan and said so. Not that it lacked substance, but it looked like most of the plans that I see in Mexico - full of activities with no focus or measures on outcomes, and so with no clear method of evaluation.

I showed him what we had achieved in Nuevo Leon and Tabasco in the nineties and explained the fundamentals of our method.

We would focus on outcomes: crime statistics. Not all crimes, only those that were significant.

We would measure and proactively publish on a monthly basis. We would use graphs, easy to read by anyone.

We needed Governor Bours to publicly announce his commitment to reduce crime by at least 25%. No politician in Mexico- or for that case, Latin America- had done this before.

We would define a statistical profile of each crime. Some crimes, like rape, assault and car theft would require a more detailed profile.

We would focus only on those municipalities and neighborhoods that comprised 80% of the problem: the Pareto Principle.

We would need to team up with many other state and municipal government offices, and particularly with the troublesome neighborhoods.

Bottom-line: we would empower the media and the people. We would flood the system with relevant information. We would open-up to constant accountability. We would put ethics in front of politics.

What were we striving for? To enhance what we call the system's preventive intelligence. Not just the police, the whole system.

Figueroa showed interest and I became very interested in his attitude.

I had given up government consultancy years ago because political leadership in Mexico is extremely scarce. Innovation, ethics and accountability are not popular words for politicians. Maintaining the status quo is more profitable for them than making a change.

-Why 25%? He asked.

-Because 25% will force us to change the system. We can try 50% if you like, its been done before, but I wouldn't advise it at this moment.

By the end of the presentation he was willing to explore. I had one condition: Governor Bours had to support the plan all the way. He agreed.

After some initial months of data searching and probing on our part, we created a set of traffic lights for each crime: green if we were within the goal, red if we were above the historical average for the specific crime, and yellow if we were in between. A format anyone could understand and evaluate. Not tables, graphs. Not absolute numbers, percentages; a right-side brain approach.

At a State Security Council meeting in March 27 of 2007, Governor Bours went public with this simple plan.

The surface looked calm, the undertow with the a radical paradigm-shift was immense:

- Politicians love to hide information unless it is good news. We would publish systematically.
- Politicians do not like to empower people. We would empower the media and the people.
- Politicians consider the media their enemy. The media was part of the team.
- Politicians love to concentrate on activity and announce spectacular strategies. We would start with sharing the problem and let everyone contribute to the solution. This was the strategy.
- Politicians avoid compromising on outcomes. We had a public goal: Easy to understand, difficult to accomplish.
- Politicians hate difficult goals. We did not know if we were capable of accomplishing it but willing to try.
- Politicians love to control. We had no control whatsoever. We did have a clear task: to find data and transform it into useful information.

The mayors of the most important cities and municipalities of Sonora (the Pareto group), accepted in public and complained in private. They had a good reason for doing so. They were to be evaluated. Sonora has a state police, but most of crime prevention and law enforcement is in the hands of the mayors.

I knew we were on the right track. Every month I facilitated the evaluation meetings in Sonora.

The first meeting would be with the state authorities. Not only the Secretary and the state police, but we also included heads of offices as diverse as Art, Sports, Education, Health, Family, Youth, etc. We would look at the crime rates - particularly those dealing with family violence, rape and assault, and make assessments and decisions. Assault is strongly related to youth gangs and requires action from many government agencies.

The next meeting was with heads of police from the Pareto municipalities. We would look at the state crime rate and then at their specific municipal crime rate. We would analyze the statistics and make decisions. This group focused on property crimes and homicide.

The next meeting was with the regional coordinators of a community program named "Pasos" (Steps), which focuses on teaming up with the neighborhoods. We would look at the crime graphics and make decisions. They were responsible for reaching the population at risk, the troublesome neighborhoods and the potential victims.

We would also meet with municipal police departments and help them out with analyzing the graphs and making decisions. Police officers tend to think reactively. Now they began to think about prevention. They tend to think it's their sole responsibility to fight crime. Now they looked at the system and began to team-up with society. They always complain about scarcity of resources. But we showed them the most valuable and free resource at hand: information. Here's an example: around 30% of car thefts are committed at shopping centers' parking lots; we published the information, the mall-managers took action, car thefts went down.

Focus, measure and decide. That was the message. Do not disperse your energy. Measure only the relevant, do not get lost in the data. You can only make three types of decisions: Go-ahead, re-think or ask for more information.

The media was fascinated with so much information. They reproduced the stats, published the preventive measures for each crime and then went on to interview some victims and some crime specialists. They had the public's attention.

We had to reach everyone; we would start with population at risk and then move on to the rest. We asked for help from all. We worked with the whole system.

At first, the quality of the decisions at all levels was poor. It takes time to understand the annual cycle of rape and the rest of the violent crimes. It takes time to make full use of the statistical profiles. Police and other government officials would look at the graphs as if it were some football game, with them as spectators. Slowly, most stepped on the field. I was trying to be the coach on the sidelines and acted as if we were a loosing team, pressing to innovate and learn from experience. The "owners" of the team, Governor Bours and Figueroa, kept both a close look and their faith. Their leadership was fundamental to sustain the effort.

The great benefit of crime rates is that they move fast. Effectiveness of decisions can be verified on a weekly or monthly basis. We wouldn't lose time with excessive analysis; we would try the hypothesis and evaluate it next month. Focus, measure, and decide; again and again, ad infinitum.

We used both crime rates and 066 (911) calls. We complemented information with surveys on rape, family violence, un-denounced crime and public perceptions. Every month more than 230 graphs appeared in the official website.

People could see what was happening at the local and state level.

Not all the mayors and heads of police understood, most resisted. We, or better yet- the system insisted. We had created a powerful model based on ethics and information. Our fulcrum was the information, our leverage, making it public.

At our meetings, it became evident who supported and who resisted, who was doing the job and who was goofing around.

And then, things started to move in the right direction. Crime rates started to fall. Family violence was the first to show green numbers. Family Violence and Rape can be decreased almost exclusively with information; crime profiles act like a mirror and thus, help potential victims take preventive measures.

We used these success stories to push on. Police departments from one city learned from experiences from another. Neighborhoods learned from each other. Newspapers learned from each other. We all learned from experimentation. They all became very creative. It was time for me to step back and learn from them.

The system has gained strength and sophistication. Information is more precise and more relevant. Not all crimes have gone down, car-theft and homicides, both related to organized crime, are in the red; some cities are trying to catch up with the leaders; mafias and drug-lords are still operating. But the pressure is on: Criminals, corrupt officials and insensitive politicians have a harder time trying to hide. The evaluation tool has to move further down the process into the prosecution and judicial authorities.

The lesson is simple: There are few exceptional politicians, the rest have to be pushed systematically by hard evidence to become exceptional.

What has been proven is that the system is working and can be extended to other states, other cities, other crimes and other social issues. In fact, the goal now is to make it a true social movement, property of the citizens.

It amazes me is that even though Sonora has become a nationwide success story or should I say, the only success story of radical crime reduction in Mexico in recent times, there has been no serious benchmarking from the national government or other state governments. The President can easily replicate success on a national level if he can put ethics in front of politics and embrace the model. He could delegate the job of building and maintaining the crime-traffic-light to independent evaluators. The data is there, hidden within his Security Council, kept in the dark, waiting.

People in Mexico are out on the streets, manifesting their despair with increasing crime rates and corrupt police, and politicians trying to handle the pressure with the old paradigms. If there is no serious effort to build an evaluation tool like Sonora did, if power is not transferred to society, the recently signed National Agreement on Security will fail and the President will face a serious political crisis.

I believe trying to control complex systems is like trying to control the weather. I believe chaos is related to force and order to power. I believe ethics reinforces life and becomes a high-energy attractor field. I believe the Universe likes truth and information. Or maybe it's just me trying to explain something I don't fully understand and just knowing that it works.

Comments on Sonora's achievements by Judge Jim Gray in his blog:

Look to Mexican example of Crime Prevention

In most jurisdictions, police are trained only to be reactive, which means that they respond to reports of crimes, investigate them, and then locate and help to prosecute the offenders. In other words, those police are "fighting crime." But with this different approach, the police were trained to think preventively, and then to share the information they received where it would do the most good, which is with the potential victims in the most vulnerable areas.

This program has demonstrably produced a material decrease in crimes, and no one can argue that this is a positive outcome. Without much difficulty, we could implement similar programs in our neighborhoods, and probably experience the same positive results.

Judge Jim Gray is well-known American jurist and the 2012 Libertarian Party vice presidential nominee. He was the presiding judge of the Superior Court of Orange County, California. He is the author of multiple books and is critical of current American drug laws.

A reverse order in science development

Its usually physicists who domain the new theories which are later introduced to the still less *predictable* sciences like biology or sociology. New rules for the inanimate world are exported to other realms of knowledge in an effort to understand the more complex world of ecosystems and society.

The Theory of Chaos is no exception and so we find economists and sociologists trying to apply the new tools to the old playground although sometimes still trying to forecast the unpredictable and therefore, not understanding the nature of chaos.

In Complex System Theory on the other hand, we find biologists making valuable breakthroughs and adding new concepts to the archives. So we have a reverse order and a major scientific paradigm-shift: New theories emerging from biology and making their way into physics.

Rules of how agents engage and relate, fractal geometry, how complexity is created from simple rules, synchronicity, self-similarity at different scales, the critical frontier between chaos and order that explains life, or how order emerges naturally without leadership are exhilarating when analyzing social systems. And yet, there is something missing, in spite of groundbreaking new theories we still don't fully grasp how to apply them to a community to understand how it behaves and more importantly, how we can influence it in a positive way; something that could help us create better social, economic and political environments.

Something is missing

But there is much more to social systems than just trying to understand them from a new scientific approach – as fresh as it may seem and as useful as it would be to include them in the social science. There are perspectives in politics, government, economics and society, which do not necessarily apply to biology: One of them is trying to make a fast and radical change in a social system, something which we have been working on since 1991.

This is the purpose of this essay: to expand the complex system theory derived from our experience of more than 18 years with crime prevention, performance management and government reform.

Our experience

We began in the early 1990s with government reform in Monterrey, Mexico. It was a fruitful time to experiment with new managerial models in the public administration. We took our theoretical baggage from Total Quality Management- TQM, which was quite popular at the time in Monterrey's prosperous industrial environment. Our purpose was to create an efficient state government. Time, resources and experience were short and the objective to change the system was extremely ambitious. The pressure was immense and this generated a very creative and practical response.

We were quite successful in spite of the odds. In a couple of years we had generated numerous success stories across government agencies and had other governments in Mexico interested in the process. We learned some of the key issues in moving a system in a radical new direction in a short time and expanded TQM theory by adding or adapting its managerial concepts to the social arena.

One of our biggest surprises was crime prevention. Crime rates plummeted in a matter of months in Monterrey. Every month we would publish the state crime reports, something quite unusual at the time in Mexico. This put pressure on police but more importantly, it also impacted on teamwork across government agencies, populations at risk and the press.

In 1998, we successfully replicated the story in the southern –and less developed- state of Tabasco with similar results. Most property crimes plummeted in the order of 20 to 50% but our biggest reward was reducing rape by more than 50% in less than a year.

In 2006, I was asked to help the State of Sonora, Mexico's second largest state, bordering with Arizona. Sonora is surrounded by high-crime states like Chihuahua to the east, Sinaloa to the south and Baja California to the west -all home to important drug-cartels. In a couple of years we had successfully reduced crime rates and established Sonora as the safest state in the US-Mexico border, including both sides of the border. It was an immense national success that contrasted with Mexico's tough moments: in 2008 people were marching in the streets protesting against insecurity in most major cities; in 2009, the crime-prevention program in Sonora won the National Award for Innovation.[1]

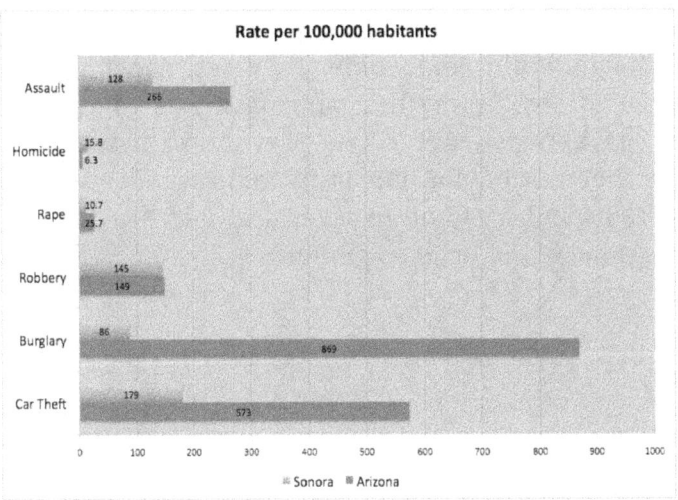

With Sonora's creativity and willingness I had honed the model and gave it a name: *Semáforo Delictivo* or *Crime Traffic Light* since we used red, yellow and green lights on the graphs to signal the monthly performance.

New Theories

Although TQM principles and tools were still used in the model I knew I was way out on uncharted territory and became intrigued about the theoretical explanation. I had induced many tools

---

[1] The details can be found in *Between Order and*

and principles by trial-and-error and knew the model worked, but I wasn't sure *why* it worked.

I started studying Chaos Theory and then moved on to Complex System Theory, a natural step. I was impressed with most of the concepts as they fitted perfectly into our findings. Later on, I looked for specific books on these two theories applied to social systems, justice and crime prevention and although I did find some, they fell short of my expectations. So my first approach was to make a bridge between our crime-prevention model and the theories of non-linear dynamics and complex systems and try it out in several conferences. These are some of the bridges I found:

Non-linear reality

One of the fundamental concepts policy-makers, analysts, public managers, the press and politicians should understand is that social systems are the most complex of all systems and therefore there is no room for linear thinking: Its is hard to pinpoint causality in complex systems, it is vain to predict precise outcomes and it is extremely dangerous to focus on simplistic solutions which may have great popular appeal to the masses but are the cause of most public policy failures.

When asked why Sonora has been successful reducing crime, people expect me to give them a precise recipe with 4 or 5 clear actions. But this perspective is wrong because it is linear: it's not a matter of technology, police control, new laws etc. We have to look at the system as a whole and what we did in Sonora was to change the system. By system we mean police, politicians, social development agencies, the mass media, and most importantly the citizens. Everyone makes better decisions in Sonora regarding crime prevention. We made a *paradigm shift* while working with the same laws, people, technology and resources. So

19

the correct answer to the question would be: We created conditions for the emergence of a new complex order.

The Universe is non-linear, we live in a non-linear reality but we still think in Newtonian terms. It's not a matter of doing A so B happens. It's much more complex than that. I'll explain this concept further when talking about fractal geometry.

Order, Chaos, Criticality

Order is contained in chaos and chaos is contained in order. It's like the yin and yang diagram.

But the terms are tricky. In non-linear dynamics chaos is considered complex order (extreme chaos is what we usually think of as chaos). It's not random and better yet, patterns emerge spontaneously in the form of attractors. An *attractor* is a set towards which a system evolves over time. An attractor can be a point, a curve or a complicated set with a fractal structure known as a *strange* attractor, the most popular being the one that looks like a butterfly.

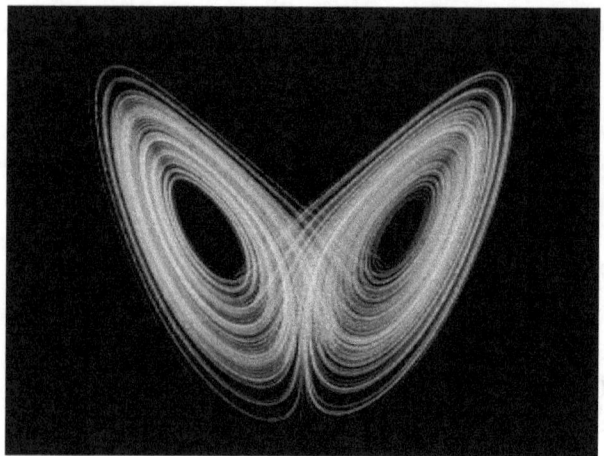

The frontier between chaos and order is *criticality* and that's the frontier that allows life to happen. Frozen or extremely chaotic realities do not support life. Life is change, constant change, but it is not random, it follows certain basic and usually simple rules. If we can understand these rules we are able to make a change in the system.

One of the key signs to observe when changing a system is oscillation. The system begins to fluctuate in what is called a bifurcation and then it does it again and again. Systems should be observed wholesomely and by results and movement and not by its parts or static dissection.

On a more practical note for policymakers I can suggest several guidelines: Turbulence is necessary and welcomed when changing a system, turbulence is part of life and some would say, it is life itself. Hierarchical control and the desire to create an extremely ordered system is highly unnatural and unfeasible, at least not without a great cost in lives, resources, creativity and success.

How does Order emerge in Social Systems?

Most importantly, complex systems have the capacity to auto-regulate themselves. Order is self-emergent.

Santiago Roel R

Fractal Geometry, Self-Similarity and the Infinite within the Finite

Fractal geometry is the geometry of nature and much more complex -and interesting- than the Euclidian geometry we learned in school. You can Google *fractals* and understand what I mean by saying simple rules create complexity and beauty in nature. It is the smart way in which nature creates. Lungs, blood vessels and most of our organs are possible because they are fractal: there is an immense surface within a limited space. The Koch Snowflake helps us understand this principle: The surrounding circle is *finite*, the surface of the snowflake can become *infinite* as new triangles are added.

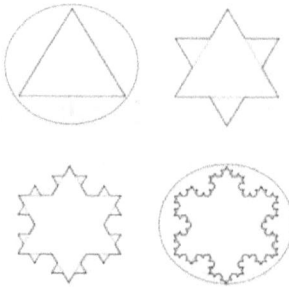

Since nature uses the same rules, there is self-similarity at different scales. Interestingly, this similarity of structure at different scales is observable in the crime graphs at the different levels of neighborhood, city, state and nation.

But let me make a bridge to something important for decision makers. What fractal geometry teaches us is that the *infinite is contained within the finite*. It is like a piano. The number of notes we can stroke is finite and contained on the keyboard, but the number of combinations for melodies and harmonies is infinite.

What does this have to do with crime prevention?

The traditional approach has been to predetermine crime prevention strategies. But the radical approach proposed here is that we can only set the basic principles, the keyboard so to speak. In fact the number of actions –decisions- taking place in the system are infinite. Every person within the system is making decisions all the time. The goal is to help the individuals within the system make better preventive decisions.

The great advantage of crime prevention is that the indicators are updated frequently and strategies can be evaluated on a weekly or monthly basis. It is unnecessary to spend much time on hypotheses; the new idea can be taken into the street and tested within a short period of time. If the new action is effective, it is reinforced; if not, it can be reformulated. This is how nature works. This process can be replicated at different scales: state, city and neighborhood. It is self-similar.

The issue then is to set the rules so the decision-making process is based on relevant information and carried out by the team. In contrast, pre-determined plans are cumbersome, expensive and most of the time, futile.

In our crime-prevention model we determine a 5-step permanent cycle: Focus, measure, communicate, make decisions and evaluate. This is how it looks:

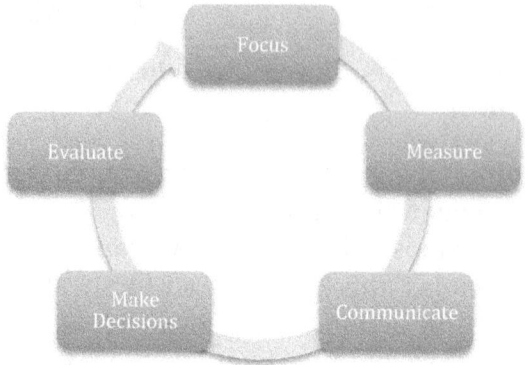

1. Focusing is extremely important. We concentrate in the most relevant crimes and neighborhoods. If measuring is involved, we use the Pareto Principle (80-20) to determine the population at risk.
2. Measure is equivalent to extracting information from the system, in this case, the system *talks* through the crime rates.
3. Communicate: We send back the information to the system to help it make better decisions.
4. Everyone makes decisions: Government, police, mass media, citizens and population

    at risk. There are three kinds of decisions: Reinforce, rethink or ask for more information.

5. Evaluate results. It is time to refocus again.

The model creates a brief monthly cycle in which order is allowed to emerge in an organic and dynamical way throughout the system.

Santiago Roel R

Emerging Order and the Control or Order Paradigm

This is one of the most astounding concepts. Order emerges naturally in complex systems. Particles, agents, animals and people follow certain basic rules within the system and create complex order; there is no hierarchy involved. This can be clearly seen in herds, flocks, swarms or crowds and can be modeled in a computer. I recommend watching *Starlings on Otmoor* in Youtube to see how thousands of birds create order and beauty without a leader.

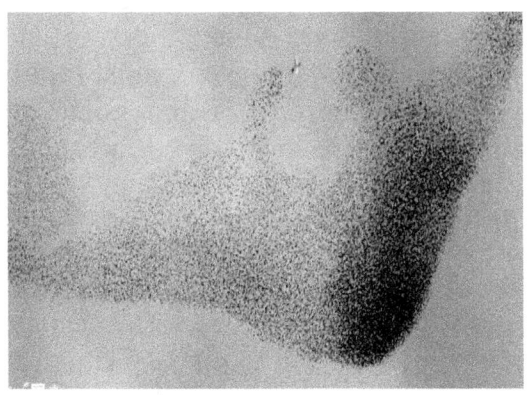

Simple rules for a crowd are: follow the person in front of you and keep a certain distance from your neighbors. We can add *avoid the predator* for herds or flocks or for potential victims in crime-prevention.

Complex systems have the ingrained capacity to auto-regulate themselves. Order emerges from chaotic scenarios and there is a huge difference between trying to regulate a system and or helping it auto-regulate itself. This is generally unknown to leaders, policy makers or analysts; most try to control everything because they truly believe this is the only path for success, when in fact, it is the contrary.

The following example is clear and radical: Most rapists are family members or friends. In Tabasco and Sonora we gave that information to potential victims and they protected themselves- radically reducing rapes. In a control approach this would have been impossible and this is what Tabasco's General Attorney believed at first. She believed her only strategy at hand was to reinforce prosecution, but this -although helpful- was a linear perspective with a small impact on the system since most rapes go unreported. Furthermore, it is impossible to protect potential victims from a control attitude: there is no way to install a policeman in every household to prevent rape from family members.

Again, the control paradigm is highly ineffective and costly because it is unnatural. The worst policies and the most counterproductive managerial, economical, social, educational or political systems come from trying to control other people. Control should be considered an extreme and temporal measure. Rules that work with the autonomy of the parts are a much better choice.

Other concepts

There are of course many other interesting concepts in Chaos and Complex Systems Theory and I highly recommend social science students, politicians trying to make a change, government officials, teachers and communicators to read as much as they can. I will include a brief list of books and articles I have found useful.

Santiago Roel R

The missing principles

The more I read about complex systems the clearer it became that something was missing. The concepts I found in the books were superb but fell short in fully explaining our success with the crime-prevention model.

Information as a _key_ element for Emerging Order

One of the concepts was information. We knew that the flow of relevant information was critical to reduce the crime rates. We empowered the system with information with the idea of raising awareness and what we call *preventive intelligence.* It worked and it worked fast.

Although I did find some interesting concepts regarding information like the rules of networks, or that turbulence or chaos are more meaningful than order, I did not find the clear concept that information flow is KEY to emerging order. It is not an additional idea it is a *fundamental* law. To reduce crime, we obtained valuable statistical information from crime reports and surveys, translated it into practical knowledge for the population and sent it back to the community in a short monthly cycle.

And so, I promptly concluded that information-flow within the system- as a fundamental principle -was the missing link to understanding emerging complex order.

We had not changed any laws in Nuevo León, Tabasco or Sonora; we did not work with police structure, equipment, technology or management; we did not try to understand the motives of crime; we didn't focus on corruption. We only focused on information: information changed police and citizens' perception and decisions, and this reduced crime.

Was this all? So I thought, but soon became proven short by a new reality.

Intention

In 2009, there were municipal and state elections in Sonora and the common turbulence that comes with them. A new party with new politicians was elected. The media pressured the new administration to continue the successful program and so they did, information kept flowing through the *Crime Traffic Light* and yet, crime started to push back and the green lights soon became red. The tools and the concepts were there but something was missing. I went back to the books. Again, I did not find a clear explanation to what was happening. So I stepped out to discover it.

What had changed in Sonora? Politicians and top level officials. What had I been working with the previous team? Paradigms. The Crime Traffic Light methodology was so simple it was misleading but the paradigm-shift underlying the method was startling and difficult to understand and sell. I can mention many of the new paradigms but the most important one was **order** versus **control**. We were not trying to control anything we were trying to help the system auto-regulate, and as you might infer this is a tough cookie to swallow for most politicians and police chiefs.

What were these new politicians missing? What had changed? **Intention**.

The method was there, the information was there, but intention had changed. The new people in charge had just stepped in from a tough political campaign and a close election. Campaigns are like a war: there are enemies, strategies, weapons, winners and losers. And so, not surprisingly, they renamed the crime-preventive program the *Great Crusade* and made a great promotional effort, but the results were partial.

The new program's objective sounded like the old one but was different in many subtle ways, and if intention changes, everything else does, even the quality and purpose of information.

I went back to the scientific books. No mention of *intention*. Okay, so this might sound unscientific, let me search for the *desired outcome* or *purpose* of the system. No mention. And then it became clear that scientists are not willing to risk mentioning a higher purpose or a desired outcome in a system - physical or biological- since this could lead to a *creationist* point of view. Although, when we talk of ecosystems there is a desired outcome: Life.

All agents, all plants and animals have an individual purpose: to live; a species purpose- to reproduce and a greater purpose: to contribute to the life-permitting conditions of the ecosystem. Scientists can debate about cooperation or competition being the main motivation, but this is secondary to the main objective: Maintaining the ecosystems conditions for life. The human body is a clear example of a higher purpose: There are 50 trillion individual cells in clear cooperation and communication for the body's well being, in alignment with a higher collective purpose and an ecosystem.

Now, when speaking of social systems we can talk about intention without having to enter a debate of a higher intelligence. Intention is very human, and so, we can move ahead. Intention is the first rule to observe for emerging order in social systems.

When I understood this, I began focusing on the paradigms behind the new administration's desire to reduce crime: Order was important, we couldn't control anything (except our own actions and even that is debatable from a psychological perspective), we had to understand and promote natural order. We were not fighting a war, there were no enemies; we did not seek revenge; we were not motivated by fear or hate: peace was a better energetic field. It was not a matter of police reaction but of social prevention. Everyone was important and if that became clear, our resources would become abundant. We had to work with the whole system: population at risk, communities. The press was part of the team, they would still criticize but now they would do it knowledgably: the quality of information would improve. No privileges for anyone: ethics above politics. These were some of the paradigms I had worked with the previous administration for more than 2 years and now I began to emphasize them with the new people in charge, not without the expected initial resistance from their part.

This may sound to some as idealistic or ethereal. I wont debate that, it is both I guess, but more importantly, it works. As intention shifted into the desired paradigms in Sonora in the first months of 2010, information and actions became aligned and crime rates began moving back into the green lights within months.

Actions followed information, information followed intention.

Santiago Roel R

Three basic principles to order complex social systems

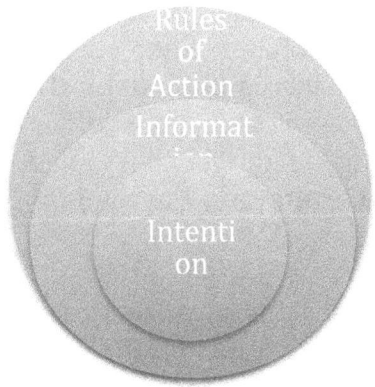

Therefore, complex social systems follow 3 fundamental laws or principles:

1. Intention
2. Information
3. Basic rules of action

Let me work my way up.

Basic Rules

Basic rules of action are given within the system: Cultural, psychological, legal, biological, economic, physical. Most scientists study these rules. In swarms, flocks or crowds, individuals attend simple rules: Follow the individual in front, keep a certain distance from the lateral companions and avoid predators or obstacles. These basic rules create order within the herd without the need of a leader. Order emerges naturally. These rules may be modeled in a computer to simulate the flock's movement with amazing results. Car traffic has similar rules and there is no need of policemen in every car to supervise and control the driver's conduct. Order emerges naturally. Physical and biological systems follow simple rules with which they create incredible sophistication and complexity. These rules apply at different scales as may be seen in fractal geometry; nature uses the same rules at every scale as building blocks for complexity.

Most politicians and leaders focus on rules attempting to change, modify or influence results by changing the law. Most theorists will analyze these rules in a traditional or unconventional way.

Chaos Theory and Complex System Theory is the *new* unconventional route by recognizing the non-linear reality of Nature. Which means the there is a complex relationship between agents that derives from a simple equation or a simple set of rules. Sensitivity to initial conditions and complexity makes it hard to make precise predictions of where the system will go or be if it is stimulated or influenced except within the range of the attractor. The system can be ordered, extremely chaotic or critical. Criticality is recognized by many as the frontier between order and extreme chaos and the place to be if you want to be alive. Life is turbulent, but not extremely turbulent and not orderly frozen. I can continue with the concepts but I am sure you get the picture of where they fit, they are the basic rules of action that individuals follow within a system and it is important to understand them in order to interact in an effective way with complex system.

Turning to social systems, we can add and analyze the economic incentives, or social rules, or psychological motivations or legal parameters, but these will only be a part of the bigger picture. In fact, while radically reducing crime in Sonora we did not go into those rules. We did not analyze why an uncle or a stepfather rapes, or why alcohol induces domestic violence, or why violent crime increases in the spring and summer, or why police corruption is needed to protect car-thefts. We did not change any law, although it would have been useful in some cases. So we can step up into the next principle: Information.

Santiago Roel R

Information

Information is key for complex systems. The birds in the flock or the wildebeest in the herd or the people in the crowd use constant and relevant information to move around. If they sense a predator or notice a slight movement in their neighbors they will react immediately. There is no leader in the flock, there is information that when applied to the rules of action helps the system auto-regulate. I have not found much literature on this topic except that turbulence has more information than order. Information within turbulence is more meaningful because something "new" is happening, there is a novel and surprising event to be noticed. On the contrary, redundancy -irrelevant, repetitive information- is high when there is no turbulence. Following a slow crowd in the street is full of redundant information, escaping a fire in a movie theatre is –on the contrary- full of relevant information.

Another interesting concept is how a network works. Not all individuals have to be linked to each other, the network communicates through nodes, this is economical and Nature is always economical. Just like your friends or the Web, there are websites or friends with many connections and other with few connections.

In our experience with crime reduction, there are practical guidelines to follow. These are some:

1) Information has to be relevant to the population at risk. People don't want to read or hear what the

government is achieving they are interested in knowing what to do to prevent emergencies or what to do in case of an emergency. They are much more interested in their immediate community and family than at a grander scale. So information is relevant if it has to do with themselves, the neighborhood, then their city, state, country and then finally, the world.

2) Information has to be friendly and easy to grasp at a blink. Graphs are better than tables. Colors are relevant. We use a right hemisphere approach: Keep it simple. Keep it fun. Keep it holistic.

3) Information has to be timely. We update information at least every week for police and at least every month for the community.

4) Women are better communicators than men. In an emergency, men fight or fly, women gather and talk. When we step into a troublesome neighborhood women are always there, caring about others and ready to understand the message and pass it on.

5) Do not fight the wave, surf it. Crowds, the press and communities focus on events. Events can create waves of information. It is of no use to waste energy trying to convince everyone that insecurity has decreased if a horrible crime makes the headlines. The crowd will follow the wave so it is useful to insert the preventive message within

the news frame. Use turbulence to reinforce the program's intention.

6) Focus. Be brief. Do not disperse. Be practical. The crowd does not have time for speeches, abstractions or rationalizations. The message has to be clear, precise, useful to the customer and clear enough for her to resend it to others. The shorter the message the more energetic and powerful it becomes.

7) Use all channels available. We send the preventive message directly to the population at risk through booklets and gatherings, but we also use the mass media, the Web, direct mail, word of mouth, the phone, SMS, classrooms, social networks, email, etc. The system has to be permeated with relevant information.

8) The strategy has to be economical in order to be sustained indefinitely.

9) Here and now. Preventive information is about the here and the now. What time of the day, what day of the week and what neighborhoods are more prone to a specific crime.

10) Focus on outputs and outcomes. Actions are only relevant in relation to outcomes. This is highly important when talking about accountability. In underdeveloped political systems accountability focuses on money spent and actions taken without concern to the most important: the outcome.

11) Empower people with the information. Fear calibrates as low energy, empowerment calibrates high.

12) There is *a tipping point*. When the preventive message reaches a critical mass the system turns in the right direction –the tipping point- and radical changes begin to happen. The idea is to create an *avalanche*. So building-up pressure is needed on a permanent basis and there is no way of predicting the time or intensity of the outcome although with experience, the tipping point can be sensed from subtle signals being sent by the system.

Intention

Rules of action and information come second to intention. Intention is much more subtle than the other 2 principles and therefore it is harder to grasp. Or sometimes, so obvious it is overseen.

Intention could be described as a purpose or a desired outcome for the system, though I prefer to think of intention as an energy field to which everything else aligns with: Information and rules of action are influenced by intention. The energy field created by intention comes first, then information and then movement or action. Intention is the most concentrated form of energy; it is the *center* of the system.

 We can easily be misguided by common slogans or objectives in public policy: "Social justice", "crime reduction", "social development" and so on. These are facades. We have to look deeper to understand intention: the real motives, beliefs, emotions and paradigms that underlie the outspoken objective.

Is there fear or hate underlying a crime program? Do we want revenge or peace?
Is there a hidden desire to control? Or a concealed need for applause and recognition? Is it for real or are we just trying to deal with the problem? Is our intention exclusive to some or inclusive to all? Are

49

we willingly or unwillingly camouflaging our intentions? More importantly, are we determined to speak and decide truthfully no matter what the truth is?

Think of your own body. Could your body survive in an untruthful environment, or with an organ trying to control or outsmart the rest, or by rebel cells concealing information?

Truthful intention is a great attractor in a complex system and although subtle and intangible, it materializes in successful outcomes like crime reduction. Successful companies, families, nations and individuals allow truth to come forward instead of projecting their own shadow on a scapegoat. Recovery programs begin with acceptance.

There are levels of awareness or conscience regarding truth. The truth of a politician is different from that of a common citizen. David R. Hawkins makes an interesting proposal in his book: Power vs. Force. He sets forth a *Map of Consciousness*, which I have found extremely interesting when relating it to intention. Most scientists will find his method of calibration highly debatable and I will not go into that, but the map is a work of art and highly insightful. Some guidelines to read the map: Log is the calibration for each Level of consciousness; Emotion is the prevailing emotion in that level. Start from Log, Level and Emotion and then take a side look to Process, Life view and God-view. The scale is logarithmic.

How does Order emerge in Social Systems?

"All levels below 200 are destructive of life in both the individual and society at large; all levels above 200 are expressions of power. The decisive level 200 is the fulcrum that divides the general areas of force and power".[2]

MAP OF CONSCIOUSNESS					
God-view	Life-view	Level	Log	Emotion	Process
Self	Is	Enlightenment	700 1000	Ineffable	Pure Consciousness
All-Being	Perfect	Peace	↑ 600	Bliss	Illumination
One	Complete	Joy	↑ 540	Serenity	Transfiguration
Loving	Benign	Love	↑ 500	Reverence	Revelation
Wise	Meaningful	Reason	↑ 400	Understanding	Abstraction
Merciful	Harmonious	Acceptance	↑ 350	Forgiveness	Transcendence
Inspiring	Hopeful	Willingness	↑ 310	Optimism	Intention
Enabling	Satisfactory	Neutrality	↑ 250	Trust	Release
Permitting	Feasible	Courage	↓ 200	Affirmation	Empowerment
Indifferent	Demanding	Pride	↓ 175	Scorn	Inflation
Vengeful	Antagonistic	Anger	↓ 150	Hate	Aggression
Denying	Disappointing	Desire	↓ 125	Craving	Enslavement
Punitive	Frightening	Fear	↓ 100	Anxiety	Withdrawal
Disdainful	Tragic	Grief	↓ 75	Regret	Despondency
Condemning	Hopeless	Apathy	↓ 50	Despair	Abdication
Vindictive	Evil	Guilt	↑ 30	Blame	Destruction
Despising	Miserable	Shame	20	Humiliation	Elimination

3

My rule of thumb is simple: If leaders are not willing to go all the way to achieve a positive outcome, if politics come before ethics and there is a desire to control, I usually step out of the program because there are no resources or strategies that can overcome this energy field.

[2] Hawkins, David R. *Power vs. Force*, USA: Hay House Inc, 2002., 76.

[3] Hawkins, David R. *Power vs. Force*, USA: Hay House Inc, 2002., 69.

Using the Hawkins scale, the crime-prevention model proposed here calibrates above 600 (Peace), while traditional crime programs usually calibrate at 150 (Hate), which reinforces aggression and revenge.

How can we change a complex social system?

1. We have to clarify the proper intention. Higher levels of conscience have better results. Let me use the Map of Consciousness: Peace, love, reason, acceptance and willingness are more powerful energy fields than: pride, anger, fear, apathy, guilt or shame. This process takes time and this is the most important step. If intention is not clear or calibrates low it is unwise to continue.

2. It is useful to analyze and list the paradigms behind the overt intention. Here is a comparison between a traditional crime program and our model:

Model Comparison

Traditional Paradigms	New Paradigms
• Reactive	• Preventive
• Disperse	• Focused
• Action/Police-oriented	• Holistic/Multidimensional/ Fractal
• Unaccountable	
• Unintentional	• Accountable
• Hierarchical/predetermined	• **Intentional**
• Secretive	• Organic: Emergent
• Control/Force	• Public
• Linear	• Order/Power
• Fear/Hate/Revenge/Desire	• Non-linear: Alive: Creative
	• Acceptance/Peace/All-Being

Regarding information, we have to obtain and constantly feedback relevant information to the system. Shorter cycles of information are more effective than longer ones. I always recommend a communication strategy to make sure relevant information is being disseminated at all scales.

If these two principles are set right, the basic rules of emerging order will operate in a positive way. Caution: it will take time for government and police to understand the whole value of intention and information.

3. Follow the 5 steps and repeat the cycle

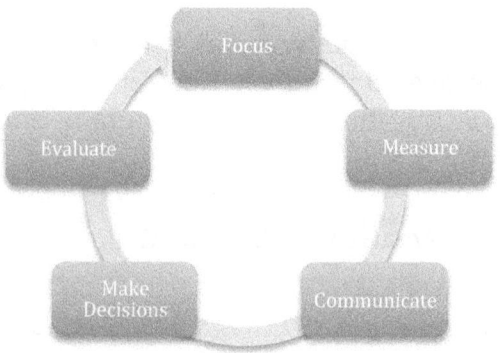

4. Create a system. The 5-step cycle has to become a system. This is the axis of transformation.

5. Once a critical mass is formed, the system will change radically in the desired direction, this is a tipping point but the new system is still fragile.

6. Insist. Learn from experience it is an ever-ending process. Life is turbulent; Order and Chaos dance together. Be creative and more importantly, allow the system to be creative.

7. Always look at the outcomes of the system this is the best measure and the best evaluation. Always take a holistic approach; do not move into details before understanding the big picture, in case of confusion, step back and take a view from the top.

8. Do not disperse. Focus. Do not waste time in sophisticated hypotheses. Be practical. Nature is economical.

The new system will emerge once the new force field is stronger than the previous one.

Paraphrasing Archimedes: To move a social system *the place to stand is* **intention** *the leverage is* **information**.

Santiago Roel R

Conclusions

Working with complex social systems has been my passion. Both government reform and crime prevention programs have showed me systems can change and they can change rapidly. Understanding how order emerges is fundamental to leaders, analysts, consultants, students and communicators. Beware of linear-thinking; most unsolved problems in our society are due to this simplistic paradigm. Chaos Theory and Complex System Theory should be part of the social sciences curricula. Focus on outcomes and always step back and move up to take a better view of the system. Do not pre-determine strategies, nature is fractal, there are infinite decisions being made within a finite framework; create the keyboard and let the melodies and harmonies emerge naturally. Weak systems seek to control others, strong systems auto-regulate themselves. Understand the power of intention and the flow of information in social systems; these principles are key for emerging order. Information is key for complex order to emerge, but information follows intention. Intention is the center, the energy field enveloping the system. Revise hidden emotions, beliefs and paradigms behind the outspoken policies or objectives. Truth is a powerful attractor for better decisions. Create a short-term cycle. Insist. Sustain a constant communication strategy this is the leverage. The avalanche, the turning point will happen sooner or later and a new complex order will manifest.

David R Hawkins' scale calibration for this book: 520

Suggested Readings / Bibliography:

1. Gladwell, Malcolm. *The Tipping Point*, New York: Bay Back Books / Little, Brown and Company, 2000.
2. Gleick, James. *Chaos: Making a New Science*. New York: Penguin Books, 2008.
3. Gribbin, John. *Deep Simplicity: Bringing Order to Chaos and Complexity*, New York: Random House, 2004.
4. Hawkins, David R. *Power vs. Force* , USA: Hay House Inc, 2002.
5. Kauffman, Stuart. *Reinventing the Sacred*. USA: Basic Books, 2008.
6. Kiel, D. & Elliot, E. *Chaos Theory in the Social Science*, USA: The University of Michigan Press, 2007.
7. Lipton, Bruce H. *Spontaneous Evolution*, USA: Hay House Inc, 2009.
8. Mitchell, Melanie. *Complexity: A Guided Tour*, New York: Oxford University Press, 2009.
9. Prigogine, Ilya. *The End of Certainty: Time, Chaos, and the New Laws of Nature.* New York: The Free Press, 1997.

10. Roel, Santiago. *Between Order and Chaos: A Mexican crime-prevention success story.* www.prominix.com, 2008.

11. Roel, Santiago. *War on Drugs: A Failed Paradigm.* www.prominix.com, 2009

12. Strogaz, Steven. *Sync: How Order Emerges from Chaos in the Universe, Nature and Daily Life,* New York: Hyperion, 2003.

13. Taylor, Marc C. *The Moment of Complexity.* London: The University of Chicago Press, 2003.

14. Waldrop, Mitchell. *Complexity: The Emerging Science at the Edge of Chaos.* New York: Touchstone, 1992.

15. Other Authors:

- Per Bak
- Bert Hellinger
- Margaret Wheatley

www.ingramcontent.com/pod-product-compliance
Lightning Source LLC
Chambersburg PA
CBHW070919180526
45168CB00005B/2076